WORLD'S FIRST ROBOT DENTIST
The Dawn of Autonomous Dental Procedures

*Exploring the Breakthroughs, Benefits, and
Ethical Implications of AI in Dentistry*

Alejandro S. Diego

Table of Contents

Introduction

In a world where technology is constantly pushing boundaries, a historic event has unfolded that promises to reshape the future of dental care. Imagine sitting in a dentist's chair, not facing the usual apprehensions but rather marveling at the precision and efficiency of a robot performing your dental procedure. This is not a scene from a science fiction movie but a groundbreaking reality brought to life by Perceptive, a Boston-based company that has successfully conducted the world's first fully autonomous dental procedure on a human patient. This monumental achievement marks a significant turning point, heralding a new era in dental care where technology and healthcare converge in ways previously thought impossible.

Perceptive, driven by visionary minds like Dr. Chris Celo, has embarked on an ambitious mission to revolutionize dental care through cutting-edge technology. Their AI-controlled robotic system, designed specifically for dental procedures, is nothing short of a marvel. Capable of preparing teeth for dental crowns with unparalleled precision, this system promises to transform the traditional dental experience. What typically requires about two hours of meticulous work by human dentists can now be completed in a mere fifteen minutes by this robotic wonder. While such a dramatic reduction in procedure time is awe-inspiring, the real magic lies in the potential for increased accuracy and efficiency, offering a glimpse into a future where dental care is faster, safer, and more accessible.

The implications of this advancement are profound. Dental procedures, often associated with anxiety and discomfort, could become more streamlined and less invasive, significantly improving the patient experience. The precision of the robotic system ensures that procedures are performed with a level of consistency and accuracy that surpasses human capabilities. This breakthrough not only enhances the quality of care but also democratizes access to high-quality dental services, potentially benefiting millions who might otherwise face barriers to receiving adequate dental care.

However, the journey to this historic milestone was far from easy. Perceptive's innovative approach combines advanced imaging technology with state-of-the-art artificial intelligence. The cornerstone of their robotic system is a handheld 3D volumetric scanner utilizing Optical Coherence Tomography (OCT). This sophisticated imaging technology creates detailed three-dimensional models of a patient's mouth, capturing not just the visible surfaces of teeth and gums but also the intricate structures beneath, such as nerves. Unlike traditional dental imaging methods like X-rays, which use ionizing radiation and provide two-dimensional images, OCT employs light beams to construct its volumetric models. This not only eliminates radiation exposure but also offers an unprecedented level of detail and accuracy.

As we delve deeper into the workings of this robotic marvel, it becomes evident that the technology's sophistication extends beyond imaging. AI algorithms analyze the 3D models to identify potential issues and formulate treatment plans. These algorithms, trained on vast datasets of dental imagery and treatment outcomes, can recognize patterns and make connections that might elude even the most experienced human dentists. The result is a harmonious blend of human expertise and artificial intelligence, working together to deliver superior dental care.

Perceptive's achievement is more than just a technological triumph; it represents a paradigm shift in how we perceive and receive dental care. The ability of the robotic system to perform procedures autonomously, even under movement-heavy conditions, showcases a level of adaptability and precision that is nothing short of revolutionary. Patients no longer need to remain perfectly still during procedures, potentially reducing discomfort and making the overall experience more pleasant.

Yet, with such transformative technology comes a host of ethical and regulatory considerations. The introduction of autonomous systems in a field as personal and delicate as dentistry raises questions about the future role of human dentists, the impact on job security, and the balance between technological efficiency and human empathy. Furthermore, extensive clinical trials and regulatory approvals will be necessary to ensure the safety and efficacy of these systems in real-world scenarios.

As we embark on this journey through the world of autonomous dental procedures, we will explore the intricacies of Perceptive's technology, the challenges and triumphs of its development, and the profound implications for the future of dental care. This book aims to provide a comprehensive understanding of how a small Boston-based company has achieved what once seemed like science fiction and what this means for the future of healthcare. Prepare to be captivated by a story of innovation, perseverance, and the relentless pursuit of excellence, all culminating in a technological breakthrough that promises to redefine the boundaries of what is possible in dental care.

Chapter 1: The Birth of Autonomous Dentistry

Perceptive, a trailblazing Boston-based company, embarked on an ambitious mission to revolutionize the field of dental care through cutting-edge technology. Founded with a vision to enhance precision, efficiency, and accessibility in dental procedures, Perceptive set out to challenge the traditional norms of dentistry. At the heart of this revolutionary endeavor is Dr. Chris Celo, the CEO and founder of Perceptive. Dr. Celo, a visionary in the field, recognized the immense potential of integrating advanced robotics and artificial intelligence into dental practices. His relentless pursuit of innovation and excellence has been the driving force behind Perceptive's groundbreaking achievements.

From its inception, Perceptive faced numerous challenges. The idea of a fully autonomous robot performing intricate dental procedures was met with skepticism and doubt. The initial phase of development involved rigorous research and development to create a robotic system capable of performing dental procedures with unprecedented precision and speed. One of the early milestones was the development of a highly advanced robotic arm equipped with various dental tools, designed to carry out procedures with a level of accuracy that surpasses human capabilities.

The journey was fraught with technical challenges, including perfecting the integration of artificial intelligence with robotic systems. Developing AI algorithms capable of analyzing complex dental images and formulating precise treatment plans was a daunting task. The team at Perceptive, led by Dr. Celo, worked tirelessly to train these algorithms using vast datasets of dental imagery and treatment outcomes. The goal was to enable the AI to recognize patterns and make connections that even experienced human dentists might overlook.

Another significant milestone was the creation of a handheld 3D volumetric scanner utilizing Optical

Coherence Tomography (OCT). This technology represented a major departure from traditional dental imaging methods. Unlike X-rays, which use ionizing radiation and provide only two-dimensional images, OCT employs light beams to construct detailed three-dimensional models of a patient's mouth. This innovation not only eliminates radiation exposure but also offers an unprecedented level of detail and accuracy, capturing both visible surfaces and underlying structures like nerves.

As the technology began to take shape, Perceptive faced the critical challenge of transitioning from controlled laboratory tests to real-world applications. Ensuring the system's reliability and safety in diverse and dynamic clinical environments required extensive testing and refinement. The team conducted numerous dry-run tests on human subjects, fine-tuning the system's ability to adapt to patient movements in real-time. This adaptability was crucial to ensure the system could operate safely and effectively even in the most movement-heavy conditions.

Despite these challenges, the perseverance and ingenuity of Dr. Celo and his team led to the successful completion of the world's first fully autonomous dental procedure on a human patient. This landmark achievement not only validated their years of hard work but also marked a significant turning point in the field of dentistry. The ability to complete a dental crown preparation in just fifteen minutes, compared to the traditional two-hour process, demonstrated the immense potential of Perceptive's technology to transform dental care.

Throughout this journey, Dr. Celo's leadership and vision have been instrumental in guiding Perceptive through the numerous obstacles and milestones. His dedication to pushing the boundaries of what is possible in dental care continues to inspire the team and drive the company's mission forward. As Perceptive moves beyond this initial success, the focus now shifts to further refining the technology, expanding its applications, and addressing the ethical and regulatory considerations that come with such groundbreaking advancements.

The inception of Perceptive, spearheaded by visionary leaders like Dr. Chris Celo, is a story of innovation, perseverance, and the relentless pursuit of excellence. The challenges faced and milestones achieved along the way have paved the path for a future where dental care is more precise, efficient, and accessible. As we continue to explore the implications of Perceptive's technology, it becomes clear that we are witnessing the dawn of a new era in dentistry, one where the fusion of robotics and artificial intelligence promises to redefine the standards of care and transform the patient experience.

Chapter 2: The Technology Behind the Magic

At the heart of Perceptive's groundbreaking achievement in autonomous dentistry is an intricate robotic system, meticulously designed to perform dental procedures with unparalleled precision and efficiency. This advanced system comprises several key components, each playing a crucial role in its overall functionality and success.

The centerpiece of the robotic system is its highly sophisticated robotic arm. This arm, equipped with various dental tools, is designed to carry out complex dental procedures with a level of precision that exceeds human capabilities. The arm's movements are controlled by advanced AI algorithms, ensuring smooth and accurate operation even in challenging conditions. Its flexibility and dexterity allow it to perform delicate tasks, such as drilling and shaping teeth for crowns, with remarkable consistency. The robotic arm's ability to adapt to real-time changes and patient movements is a testament to the sophisticated engineering and design behind it.

Complementing the robotic arm is the system's advanced imaging technology, which is fundamental to its success. Perceptive has developed a state-of-the-art handheld 3D volumetric scanner that utilizes Optical Coherence Tomography (OCT). This scanner is capable of creating highly detailed three-dimensional models of a patient's mouth, capturing not only the visible surfaces of teeth and gums but also the intricate structures beneath, such as nerves and bone.

OCT represents a significant departure from traditional dental imaging methods. Unlike X-rays, which use ionizing radiation to produce two-dimensional images, OCT employs light beams to construct volumetric models. This light-based approach eliminates the risks associated with radiation exposure, making it a safer option for patients. Additionally, OCT provides a level of detail and accuracy that far surpasses that of traditional

X-rays. The three-dimensional models generated by OCT offer a comprehensive view of the dental structures, enabling more precise diagnosis and treatment planning.

One of the major advantages of OCT over traditional imaging methods is its ability to capture fine details. X-rays, while useful for detecting cavities and bone issues, often fall short when it comes to visualizing soft tissues and fine anatomical structures. OCT, on the other hand, can reveal the minutiae of dental anatomy, such as the exact location of nerves and the structure of the soft tissues. This level of detail is crucial for planning complex dental procedures and ensuring their success.

The handheld 3D volumetric scanner, powered by OCT, plays a pivotal role in the workflow of Perceptive's robotic system. The process begins with the scanner capturing a detailed 3D model of the patient's mouth. This model serves as the foundation for the AI algorithms that analyze the data to identify potential issues and formulate precise treatment plans. These algorithms, trained on extensive datasets of dental imagery and outcomes, are capable of recognizing patterns and making connections that might elude even the most experienced human dentists.

Once the treatment plan is finalized, the robotic arm takes over to execute the procedure. The arm's tools are designed to perform specific tasks with high precision, ensuring that each step of the procedure is carried out accurately. For example, during the preparation of a tooth for a crown, the robotic arm can drill and shape the tooth with a level of consistency and accuracy that minimizes the risk of errors and complications.

The system's ability to adapt to patient movements is another remarkable feature. Traditional dental procedures often require patients to remain perfectly still, which can be uncomfortable and challenging, especially for longer procedures. The robotic system developed by Perceptive can adjust its actions in real-time based on the patient's movements, reducing the need for complete stillness and enhancing the overall patient experience.

In addition to the technical components, the integration of artificial intelligence is what truly sets Perceptive's system apart. The AI-driven decision-making process involves analyzing the 3D models to detect cavities, identify areas that need treatment, and determine the best course of action. This AI capability not only improves the accuracy of diagnoses but also streamlines the treatment process, making it faster and more efficient.

The differences between OCT and traditional dental imaging methods highlight the innovative nature of Perceptive's technology. Traditional X-rays, while effective for certain diagnoses, expose patients to ionizing radiation and provide limited views of the dental structures. OCT, with its use of light beams, eliminates radiation risks and offers comprehensive, detailed views of both hard and soft tissues. This advancement in imaging technology is a cornerstone of the robotic system's ability to perform precise and efficient dental procedures.

Perceptive's robotic system represents a fusion of advanced engineering, cutting-edge imaging technology, and sophisticated artificial intelligence. Each component, from the robotic arm and its tools to the handheld 3D volumetric scanner and OCT, plays a vital role in delivering a new standard of care in dentistry. By combining these elements, Perceptive has created a system that not only enhances the precision and efficiency of dental procedures but also transforms the patient experience, making dental care safer, more comfortable, and more accessible.

Chapter 3: AI-Driven Decision Making

The creation of 3D models and the development of AI algorithms are at the heart of Perceptive's revolutionary robotic dental system. These advanced technologies work in tandem to transform traditional dental procedures, making them more precise, efficient, and accessible. The journey begins with the creation of detailed 3D models, which serve as the foundation for the AI-driven analysis and treatment planning.

Using the handheld 3D volumetric scanner equipped with Optical Coherence Tomography (OCT), Perceptive's system captures highly detailed images of a patient's mouth. Unlike traditional dental imaging methods, OCT employs light beams to construct volumetric models without the use of ionizing radiation. This technology not only ensures patient safety but also provides an unprecedented level of detail, capturing both the visible surfaces of teeth and gums and the intricate structures beneath, such as nerves and bone. These comprehensive 3D models are critical for accurate diagnosis and treatment planning.

Once the 3D model is created, the next step involves the application of sophisticated AI algorithms. These algorithms are designed to analyze the detailed images and identify potential dental issues with remarkable accuracy. The AI system is built on machine learning techniques, trained on extensive datasets of dental imagery and treatment outcomes. This training allows the AI to recognize patterns and anomalies that may not be immediately apparent to even the most experienced human dentists.

The process of data analysis begins with the AI examining the 3D model to detect common dental problems such as cavities, gum disease, and structural abnormalities. By comparing the patient's dental anatomy with the vast repository of data it has learned from, the AI can identify issues with high precision. For example, it can detect cavities at an early stage by analyzing subtle changes in the tooth structure that might be missed in traditional two-dimensional X-rays.

After identifying the dental issues, the AI proceeds to formulate a comprehensive treatment plan. This involves determining the best course of action to address the identified problems. The AI considers various factors, including the severity of the issues, the overall health of the patient's mouth, and the most effective treatment methods. By leveraging its extensive knowledge base, the AI can propose treatment plans that are not only precise but also tailored to the individual needs of each patient.

One of the most significant advantages of using AI in treatment planning is its ability to integrate multiple data points and provide a holistic view of the patient's dental health. The AI can cross-reference current dental conditions with historical data, predicting potential future issues and recommending preventative measures. This proactive approach ensures that patients receive comprehensive care, addressing not only the immediate problems but also safeguarding their long-term dental health.

Despite the sophistication of the AI, human dentists remain integral to the process. The AI's analysis and proposed treatment plans are reviewed by a human dentist, who brings invaluable expertise and judgment to the table. This collaboration between AI and human dentists ensures that the final treatment plan is both clinically sound and tailored to the specific needs of the patient. The human dentist's role is crucial in interpreting the AI's recommendations, making nuanced decisions based on their professional experience, and providing the empathetic care that patients require.

The human-in-the-loop approach, where AI and human dentists work together, represents a harmonious blend of technology and human expertise. The AI handles the data-intensive tasks, providing detailed analyses and treatment options, while the human dentist oversees the process, making final decisions and ensuring the highest standards of patient care. This collaborative model enhances the efficiency and accuracy of dental procedures, allowing dentists to focus more on patient interaction and less on routine diagnostic tasks.

Once the treatment plan is approved, the robotic system takes over to execute the procedure. The robotic arm, guided by the AI's precise instructions, performs the necessary dental work with unmatched accuracy. Whether it involves drilling and shaping a tooth for a crown or performing other intricate tasks, the robotic arm's precision ensures consistent and high-quality results. The ability of the robotic system to adapt to real-time changes and patient movements further enhances the safety and effectiveness of the procedures.

In conclusion, the integration of 3D modeling, AI algorithms, and robotic execution represents a significant advancement in dental care. By creating detailed 3D models using OCT and leveraging AI for data analysis and treatment planning, Perceptive's system offers a new level of precision and efficiency. The collaboration between AI and human dentists ensures that patients receive comprehensive and personalized care, combining the best of technology and human expertise. This innovative approach not only transforms traditional dental procedures but also sets a new standard for the future of dentistry, making high-quality dental care more accessible and effective for all.

Chapter 4: Precision and Efficiency

Perceptive's revolutionary robotic system is built around a highly sophisticated robotic arm that has the potential to redefine dental procedures with its capabilities and precision. This robotic arm, meticulously engineered and equipped with various dental tools, is designed to perform intricate tasks with a level of accuracy and consistency that surpasses human capabilities. Its advanced features enable it to deliver high-quality dental care, revolutionizing the way dental procedures are performed.

The robotic arm's capabilities extend far beyond basic dental tasks. It is designed to carry out complex procedures, such as preparing teeth for dental crowns, with remarkable precision. The arm's movements are controlled by advanced AI algorithms, which ensure smooth and accurate operations even in the most challenging conditions. This precision is crucial for procedures that require meticulous attention to detail, such as drilling and shaping teeth. The robotic arm can execute these tasks with a degree of exactness that minimizes the risk of errors and complications, ensuring superior outcomes for patients.

Time efficiency is another significant advantage of Perceptive's robotic system. Traditional dental procedures, especially those involving crown preparation, can be time-consuming, often requiring about two hours and multiple visits to complete. In stark contrast, Perceptive's robotic system can perform the same task in approximately fifteen minutes. This dramatic reduction in procedure time is not only impressive but also has profound implications for patient care. Shorter procedures mean less time in the dental chair, reducing patient discomfort and anxiety. Additionally, the increased efficiency allows dental practices to serve more patients in less time, improving overall productivity and accessibility of care.

The adaptability of the robotic system to patient movements is one of its most groundbreaking features. Traditional dental procedures often require patients to remain perfectly still, which can be challenging, especially for longer procedures. Perceptive's robotic arm, however, is designed to adapt to real-time changes and patient movements, ensuring safe and effective operations even when patients are unable to remain completely still. This adaptability is achieved through continuous monitoring and adjustment of the arm's movements, guided by the AI algorithms. The ability to adapt in real-time enhances the patient experience by reducing the need for rigid stillness, making procedures more comfortable and less stressful.

One of the key benefits of the robotic system is its potential for delivering consistent results. Human performance can vary due to fatigue, stress, or other factors, which can impact the quality of dental procedures. The robotic arm, however, operates with unwavering precision, ensuring consistent outcomes regardless of external conditions. This consistency is particularly important for procedures that require a high degree of accuracy, such as dental crown preparations. By eliminating the variability associated with human performance, the robotic system ensures that each procedure is performed to the highest standard, improving overall patient outcomes.

Moreover, the reduced procedure times facilitated by the robotic system have significant implications for dental practices. By completing procedures more quickly and efficiently, dental professionals can increase the number of patients they see in a day, thereby enhancing the accessibility of dental care. This increased efficiency not only benefits patients by reducing wait times but also enables dental practices to operate more effectively and profitably.

The potential for reduced procedure times also extends to the patient experience. Shorter dental visits mean less time spent in the chair, which can be a major source of anxiety for many patients. The robotic system's ability to perform tasks quickly and accurately can make dental visits more pleasant and less daunting, encouraging more people to seek regular dental care and maintain better oral health.

In conclusion, Perceptive's robotic arm represents a significant advancement in dental technology, offering capabilities and precision that far exceed traditional methods. Its ability to perform complex procedures with high accuracy, adapt to patient movements in real-time, and deliver consistent results makes it a game-changer in the field of dentistry. The time efficiency of the robotic system, compared to human procedures, highlights its potential to transform dental practices by increasing productivity and accessibility of care. As we continue to explore the full potential of this innovative technology, it is clear that Perceptive's robotic system is poised to set a new standard in dental care, making high-quality, efficient, and comfortable dental treatments more accessible to all.

Chapter 5: Clinical Trials and Real-World Application

The journey of Perceptive's robotic dental system from development to real-world application has been marked by rigorous testing and evaluation. Initial trials on human subjects have provided valuable insights into the system's performance, highlighting its potential benefits while also identifying the challenges that need to be addressed as the technology moves closer to widespread adoption.

The initial trials were conducted to test the safety, accuracy, and efficiency of the robotic system in performing dental procedures, specifically focusing on the preparation of teeth for crowns. These trials involved a diverse group of patients, each presenting unique anatomical and behavioral characteristics. The results from these trials were promising, showcasing the system's ability to perform complex dental tasks with a high degree of precision and significantly reduced procedure times.

One of the key findings from the trials was the system's exceptional accuracy. The robotic arm, guided by advanced AI algorithms and detailed 3D models created using Optical Coherence Tomography (OCT), was able to drill and shape teeth with a level of precision that surpassed traditional methods. This accuracy is critical in ensuring that the dental crowns fit perfectly, reducing the risk of complications and improving the overall success rate of the procedures.

The system's performance in diverse scenarios was also evaluated during the trials. Patients with varying levels of dental complexity and different degrees of movement were included to test the adaptability of the robotic arm. The system demonstrated an impressive ability to adapt to real-time changes, making continuous adjustments based on the patient's movements. This adaptability is particularly important in real-world settings where maintaining perfect stillness is often challenging for patients. The ability to perform precise dental work even under movement-heavy conditions showcases the system's robustness and potential for broader application.

Clinical trials also highlighted several significant benefits of the robotic system. One of the most notable advantages was the reduction in procedure times. Traditional crown preparation typically takes about two hours and may require multiple visits. In contrast, the robotic system was able to complete the same procedure in approximately fifteen minutes. This dramatic reduction in time not only enhances the patient experience by minimizing discomfort and anxiety but also allows dental practices to increase their throughput, serving more patients in less time.

Another benefit observed during the trials was the consistency of results. Unlike human dentists, whose performance can be affected by factors such as fatigue or stress, the robotic system maintained a high level of precision and consistency throughout the procedures. This reliability ensures that each patient receives the same high standard of care, improving overall treatment outcomes.

However, the trials also revealed several challenges that need to be addressed as the system transitions from controlled testing environments to real-world applications. One of the primary challenges is the variability in patient anatomy and behavior. While the robotic system performed well under controlled conditions, the diversity of real-world scenarios presents additional complexities. Each patient's dental anatomy is unique, and factors such as unexpected movements or variations in dental structures can impact the system's performance. Ensuring that the system can reliably handle these variations will be crucial for its success in everyday dental practices.

Another challenge is the integration of the robotic system into existing dental workflows. Traditional dental procedures rely heavily on the expertise and judgment of human dentists. Transitioning to a system where a robot performs significant portions of the work requires changes in how dental practices operate. Dentists need to be trained to work with the robotic system, and workflows must be adapted to incorporate this new technology. This integration process can be complex and time-consuming, requiring significant investment in training and infrastructure.

Patient acceptance is another critical factor. While the benefits of reduced procedure times and increased precision are clear, some patients may be apprehensive about having a robot perform their dental work. Building trust and confidence in the technology will be essential for its widespread adoption. This can be achieved through transparent communication about the safety, efficacy, and benefits of the system, as well as positive patient experiences during initial deployments.

Regulatory approval is another significant hurdle. Before the robotic system can be widely adopted, it must undergo rigorous testing and approval by regulatory bodies such as the Food and Drug Administration (FDA) in the United States. This process ensures that the technology meets stringent safety and efficacy standards. While Perceptive has made significant progress, obtaining regulatory approval is a complex and lengthy process that requires thorough documentation and extensive clinical evidence.

In conclusion, the initial trials of Perceptive's robotic dental system on human subjects have demonstrated its potential to revolutionize dental care with exceptional accuracy, reduced procedure times, and consistent results. However, the transition to real-world applications presents several challenges, including variability in patient anatomy and behavior, integration into existing workflows, patient acceptance, and regulatory approval. Addressing these challenges will be critical for the successful adoption of this groundbreaking technology. As Perceptive continues to refine and enhance its system, the future of dental care looks promising, with the potential to offer high-quality, efficient, and comfortable treatments to patients around the world.

Chapter 6: Ethical and Regulatory Considerations

The advent of autonomous dental procedures heralds a new era of precision and efficiency in dental care, but it also brings with it a host of ethical implications that must be carefully considered. The integration of advanced robotics and AI into such a personal and delicate field as dentistry raises important questions about the role of human practitioners, patient trust, and the potential for misuse of the technology.

One of the primary ethical concerns revolves around the potential displacement of human dentists. While the technology aims to augment rather than replace human practitioners, the automation of routine dental procedures could lead to a significant shift in the role of dental professionals. Dentists may find themselves transitioning from direct care providers to supervisors of AI systems. This shift could impact the dynamics of the profession, potentially reducing the need for human expertise in routine procedures but increasing the need for advanced training in managing and interpreting AI-driven systems.

Another ethical issue is the potential for voice spoofing and impersonation. Advanced AI systems like Perceptive's robotic dental technology rely heavily on voice commands and interactions. This opens up the possibility of malicious actors using AI to spoof voices or impersonate individuals, leading to fraudulent activities or breaches of patient confidentiality. Ensuring the security of these systems and developing robust authentication protocols will be crucial to prevent such abuses.

The path to regulatory approval for autonomous dental systems like Perceptive's robotic arm involves navigating a complex landscape of safety and efficacy standards. In the United States, the Food and Drug Administration (FDA) plays a critical role in evaluating new medical technologies. To obtain FDA approval, Perceptive must conduct extensive clinical trials and provide comprehensive data demonstrating the safety and effectiveness of their system. This process is rigorous and time-consuming, designed to ensure that the technology does not pose any undue risk to patients and that it performs as intended in real-world conditions.

The regulatory challenges are not limited to technical performance alone. Ethical considerations, such as patient consent, data privacy, and the transparency of AI decision-making processes, are also scrutinized. Regulatory bodies will require assurance that patients are fully informed about the use of autonomous systems in their care and that their personal data is protected. Moreover, the AI algorithms used in the system must be transparent and explainable, ensuring that human practitioners can understand and trust the recommendations made by the AI.

In light of these challenges and the potential for misuse, companies like Microsoft, which are at the forefront of developing advanced AI technologies, have taken a cautious approach. Microsoft's decision to keep Vall-E2, their advanced AI speech tool, under wraps reflects a responsible and ethical stance on the deployment of powerful AI technologies. By refraining from releasing Vall-E2 to the public, Microsoft aims to prevent its misuse in activities such as voice spoofing, impersonation, and other malicious acts. This decision underscores the importance of addressing ethical and security concerns before widely deploying such technologies.

Microsoft's approach highlights the need for a balanced view of technological advancements. While the potential benefits of AI and robotics in fields like dentistry are immense, it is equally important to mitigate risks and ensure that the technology is used responsibly. This includes ongoing research and development to improve security measures, continuous monitoring of the technology's impact, and collaboration with regulatory bodies to establish clear guidelines and standards.

The ethical implications of autonomous dental procedures extend beyond technical and regulatory concerns. They also encompass broader societal issues, such as ensuring equitable access to advanced dental care. As these technologies become more prevalent, it is essential to address disparities in access to ensure that all patients, regardless of their socioeconomic status, can benefit from the advancements in dental care. This requires concerted efforts from policymakers, healthcare providers, and technology developers to make advanced dental technologies affordable and accessible to a wider population.

In conclusion, while the development of autonomous dental procedures represents a significant advancement in dental care, it is accompanied by a range of ethical implications that must be carefully addressed. The potential risks of voice spoofing and impersonation, the rigorous regulatory hurdles, and the need for responsible deployment of AI technologies underscore the complexity of integrating such innovations into healthcare. Companies like Microsoft, by taking a cautious and ethical approach, set a positive example for the industry. As we move forward, balancing the promise of technological advancements with ethical responsibility will be crucial to ensuring that these innovations truly benefit society and enhance patient care.

Chapter 7: Impact on the Dental Profession

The integration of automation and AI into dentistry is poised to bring about profound changes in the role of dentists, transforming not only how they practice but also how they are trained and educated. As dental technology continues to advance, the relationship between AI and human judgment will evolve, necessitating a shift in the traditional roles and responsibilities of dental professionals.

Automation in dentistry, exemplified by systems like Perceptive's robotic dental arm, will likely redefine the role of dentists by shifting their focus from performing routine procedures to overseeing and managing advanced AI-driven technologies. Dentists will increasingly become supervisors and coordinators, ensuring that the AI systems are functioning correctly and making appropriate decisions based on the data they analyze. This transition from direct care providers to technology managers will require dentists to develop new skills and competencies, particularly in the areas of AI and robotics.

The increased use of AI and automation in dentistry will also demand changes in the education and training of future dental professionals. Traditional dental curricula, which primarily focus on manual skills and clinical knowledge, will need to be updated to include training in advanced technologies. Dental students will need to learn how to operate and interact with AI systems, understand the principles of machine learning and data analysis, and develop skills in troubleshooting and maintaining robotic equipment.

In addition to technical skills, future dental education will need to emphasize the importance of integrating human judgment with AI recommendations. While AI systems can analyze vast amounts of data and identify patterns that may not be immediately apparent to human practitioners, the final decision-making process will still rely heavily on human judgment. Dentists will need to interpret AI-generated data, consider the patient's overall health and individual circumstances, and apply their clinical expertise to make informed treatment decisions.

To prepare dental professionals for these changes, dental schools will need to develop comprehensive training programs that blend traditional dental education with advanced technological instruction. This might include courses on AI and robotics, hands-on training with dental automation systems, and interdisciplinary collaboration with fields such as computer science and engineering. By equipping future dentists with a broad skill set that encompasses both clinical and technological knowledge, dental education can ensure that practitioners are well-prepared for the evolving landscape of dental care.

The evolving relationship between AI and human judgment in dentistry represents a significant shift in the practice of dental care. AI systems, with their ability to analyze complex data and provide precise recommendations, offer valuable support to human dentists. However, the role of human judgment remains crucial in interpreting these recommendations and making final decisions. This collaborative relationship between AI and human practitioners enhances the accuracy and efficiency of dental procedures, leading to better patient outcomes.

One of the key benefits of AI in dentistry is its ability to handle routine and repetitive tasks, freeing up dentists to focus on more complex and nuanced aspects of patient care. For example, while the AI system might perform the initial diagnosis and treatment planning, the dentist can concentrate on patient communication, providing personalized care, and addressing any concerns or questions the patient might have. This division of labor allows dentists to leverage the strengths of AI while maintaining the human touch that is essential in healthcare.

Furthermore, the integration of AI into dentistry can enhance the diagnostic process by providing dentists with detailed insights and recommendations based on comprehensive data analysis. AI algorithms can identify patterns and anomalies in dental images that might be missed by human eyes, leading to earlier detection of dental issues and more accurate treatment plans. This collaboration between AI and human judgment ensures that patients receive the best possible care, combining the precision of technology with the expertise and empathy of human practitioners.

As the relationship between AI and human judgment in dentistry continues to evolve, it will be important to establish clear guidelines and protocols to ensure that the use of AI is ethical and effective. Dentists will need to remain vigilant in monitoring the performance of AI systems, ensuring that they are used appropriately and that patient safety is always prioritized. Additionally, ongoing research and development will be necessary to continually improve the capabilities of AI systems and to address any emerging challenges or concerns.

In conclusion, the integration of automation and AI into dentistry is set to transform the role of dentists, requiring them to develop new skills and competencies while continuing to rely on their clinical expertise and human judgment. Future dental education will need to adapt to these changes, providing comprehensive training in advanced technologies alongside traditional dental instruction. The evolving relationship between AI and human judgment will enhance the accuracy and efficiency of dental care, leading to better patient outcomes and a more effective healthcare system. As we embrace these advancements, it is essential to balance the benefits of technology with the irreplaceable value of human care, ensuring that the future of dentistry is both innovative and compassionate.

Chapter 8: Patient Perspectives and Acceptance

The integration of advanced robotics and AI in dental care offers numerous potential benefits for patients, significantly enhancing their overall experience and treatment outcomes. These benefits range from reduced procedure times and increased precision to improved accessibility and cost-effectiveness of dental care.

One of the most immediate and noticeable benefits for patients is the reduction in procedure times. Traditional dental procedures, particularly complex ones like preparing teeth for crowns, can be time-consuming and often require multiple visits. With Perceptive's robotic system, these procedures can be completed in a fraction of the time. For instance, a process that typically takes about two hours can now be accomplished in approximately fifteen minutes. This dramatic reduction in time not only minimizes the duration of discomfort and anxiety associated with dental visits but also allows patients to return to their daily activities much sooner. Shorter procedure times are particularly advantageous for patients with busy schedules or those who experience significant anxiety during dental visits.

Increased precision is another critical benefit of robotic dental procedures. The advanced imaging technology and AI-driven analysis enable the robotic arm to perform tasks with a level of accuracy that surpasses human capabilities. This precision is essential for procedures that require meticulous attention to detail, such as drilling and shaping teeth for crowns. The ability to execute these tasks with such accuracy reduces the risk of errors and complications, leading to better long-term outcomes for patients. Furthermore, the consistency of robotic systems ensures that each procedure is performed to the highest standard, enhancing the overall quality of care.

However, the introduction of robotic dental procedures also brings forth patient concerns and anxieties that need to be addressed. Many patients may feel uneasy about the idea of a robot performing dental work, fearing the loss of human touch and the potential for technical malfunctions. To alleviate these concerns, it is crucial to provide patients with clear and transparent information about the technology, its safety features, and the benefits it offers. Educating patients on how the robotic system operates, including the role of the human dentist in overseeing and validating the procedure, can help build trust and confidence in the technology.

Ensuring that patients understand that the robotic system is designed to work in collaboration with human dentists is vital. The AI and robotic arm handle the data-intensive tasks and precise execution, while the human dentist remains integral to the process, making final decisions and providing personalized care. Highlighting this collaborative approach can reassure patients that they are still receiving the empathetic and expert care of a human professional, enhanced by the precision and efficiency of advanced technology.

The impact on patient experience and satisfaction is expected to be significant. Reduced procedure times and increased precision contribute to a more comfortable and efficient dental visit, which can enhance overall patient satisfaction. The ability of the robotic system to adapt to patient movements in real-time further improves the experience by reducing the need for patients to remain perfectly still, making procedures more comfortable and less stressful. Positive experiences with the technology are likely to lead to higher patient satisfaction and potentially greater acceptance of robotic dental procedures in the future.

Cost implications and accessibility are critical factors in the widespread adoption of advanced dental technologies. While the initial investment in robotic systems can be substantial, the increased efficiency and precision can lead to cost savings over time. Reduced procedure times mean that dental practices can serve more patients, potentially lowering the cost per procedure. Additionally, the consistency and accuracy of robotic systems can decrease the likelihood of complications and follow-up treatments, further reducing overall costs.

In terms of accessibility, the efficiency of robotic systems can help address shortages of dental professionals, particularly in underserved areas. By enabling dentists to perform more procedures in less time, robotic systems can increase the availability of high-quality dental care to a broader population. This democratization of dental care ensures that more patients have access to advanced treatments, regardless of their geographical location or socioeconomic status.

In conclusion, the integration of robotics and AI in dental care offers substantial benefits for patients, including reduced procedure times, increased precision, and improved overall experience and satisfaction. Addressing patient concerns through education and transparent communication is essential for building trust in the technology. The impact on cost and accessibility also holds promise for making advanced dental care more widely available and affordable. As these technologies continue to evolve and become more integrated into dental practices, they have the potential to transform the field, providing patients with high-quality, efficient, and accessible care.

Chapter 9: The Future of Dental Automation

The future of dental care, bolstered by ongoing advancements in robotics and AI, promises a landscape of continuous innovation and transformation. The potential future advancements and new procedures, the vision of fully automated dental clinics, the long-term impact on the dental profession and healthcare system, and the public's acceptance and adaptation to these technologies all paint an exciting and dynamic picture of what lies ahead.

As technology evolves, we can anticipate several significant advancements in autonomous dental procedures. Beyond crown preparations, future developments might include more complex and varied treatments such as fillings, root canals, and orthodontic adjustments. Innovations in AI and machine learning could further enhance the precision and accuracy of these procedures, making it possible to perform intricate dental surgeries that require a high degree of skill and consistency. Additionally, improvements in imaging technology might enable real-time, high-resolution scans that provide even more detailed insights into dental health, allowing for more effective diagnosis and treatment planning.

The vision of fully automated dental clinics represents the pinnacle of these technological advancements. In such clinics, a range of dental procedures could be performed autonomously by robotic systems under the supervision of human dentists. These clinics would be equipped with advanced diagnostic tools, robotic arms capable of performing a variety of procedures, and integrated AI systems that manage patient data and treatment plans. The seamless integration of these technologies would not only streamline dental care but also significantly reduce the time and resources required for treatment.

In a fully automated dental clinic, patients might experience a more efficient workflow from the moment they enter. Automated check-in systems could streamline administrative tasks, while AI-driven diagnostics could quickly assess the patient's needs and recommend treatment options. Robotic systems would then execute the necessary procedures with unparalleled precision, all under the watchful eye of a human dentist who ensures quality and safety. This vision of the future holds the promise of making dental care more accessible, efficient, and consistent.

The long-term impact of these advancements on the dental profession and the broader healthcare system will be profound. For dental professionals, the role will evolve from performing routine procedures to focusing on more complex and nuanced aspects of patient care. Dentists will become supervisors of advanced technologies, interpreting AI-driven data, making critical decisions based on comprehensive analyses, and providing the human touch that is essential in healthcare. This shift will require significant changes in dental education and training, preparing future dentists to work alongside and manage these advanced systems effectively.

The broader healthcare system will also benefit from these advancements. Automated dental clinics could alleviate the strain on traditional dental practices, particularly in areas with a shortage of dental professionals. By increasing the efficiency and capacity of dental care providers, these technologies could help meet the growing demand for dental services. Furthermore, the standardization and consistency offered by robotic systems could improve overall treatment outcomes, leading to better oral health for the population at large.

Public acceptance and adaptation to autonomous dental technologies will be crucial for their successful implementation. While the benefits of reduced procedure times, increased precision, and improved patient experience are clear, gaining public trust in these technologies will require addressing concerns and anxieties. Effective communication about the safety, efficacy, and advantages of robotic dental procedures will be essential. Demonstrating positive outcomes through transparent clinical trials and sharing success stories can help build confidence in these innovations.

Patient education will play a significant role in fostering acceptance. Clear explanations of how the technology works, the role of human oversight, and the benefits to patient care can alleviate fears and misconceptions. Additionally, offering patients the opportunity to experience the technology in a supportive and reassuring environment can help them become more comfortable with the idea of robotic-assisted dental care.

Adapting to these changes will also involve regulatory and ethical considerations. Ensuring that the deployment of autonomous dental technologies is conducted responsibly, with stringent safeguards to protect patient data and privacy, will be essential.

Regulatory bodies will need to develop frameworks that address the unique challenges posed by these technologies, ensuring that they meet high standards of safety and efficacy.

In conclusion, the future of dental care is poised for remarkable advancements driven by robotics and AI. The potential for new procedures, the vision of fully automated dental clinics, and the transformative impact on the dental profession and healthcare system present exciting opportunities. As public acceptance grows and regulatory frameworks evolve, these technologies will become integral to delivering high-quality, efficient, and accessible dental care. The journey ahead will require collaboration, innovation, and a commitment to balancing technological progress with ethical responsibility, ensuring that the benefits of these advancements are realized for all.

Conclusion

The journey through the development and implications of the world's first fully autonomous robotic dentist has highlighted both the incredible potential and the significant challenges of integrating advanced AI and robotics into dental care. Perceptive's groundbreaking technology has set a new standard for precision and efficiency in dental procedures, transforming how dental care is delivered and experienced.

Throughout this book, we explored the inception of Perceptive and the vision of pioneers like Dr. Chris Celo. The creation of detailed 3D models using Optical Coherence Tomography (OCT) and the development of sophisticated AI algorithms were pivotal in achieving unprecedented accuracy and efficiency in dental procedures. The collaborative relationship between AI systems and human dentists was emphasized, showcasing how technology can enhance the expertise and care provided by human professionals.

Clinical trials demonstrated the practical benefits of this technology, including significantly reduced procedure times and increased precision. Patients experienced less discomfort and improved outcomes, while the system's adaptability to real-time patient movements enhanced the overall experience and satisfaction.

However, the integration of such advanced technology also brings forth important ethical considerations. The potential displacement of human dentists, risks associated with voice spoofing and impersonation, and the need for robust regulatory oversight are critical issues that must be addressed. Ensuring that these technologies are implemented responsibly requires careful consideration of these ethical implications.

The future of dental education and training will need to adapt to these technological advancements, equipping dental professionals with the skills to operate and manage advanced AI-driven systems. The evolving role of dentists will likely involve more oversight and complex decision-making, ensuring that the human touch remains integral to patient care.

Public acceptance of these technologies is crucial for their successful implementation. Transparent communication about the safety, efficacy, and benefits of robotic dental systems will be essential in building trust. Addressing patient concerns and providing positive experiences will foster wider acceptance and integration of these technologies into everyday dental care.

Looking ahead, the potential for new procedures and fully automated dental clinics is both exciting and transformative. These advancements promise to alleviate the burden on traditional dental practices and improve access to high-quality care. However, balancing these technological advancements with ethical responsibility is paramount. Ensuring that innovations enhance patient care without compromising ethical standards will be a critical challenge.

In conclusion, the advent of autonomous dental procedures marks a significant milestone in dental care. The benefits of reduced procedure times, increased precision, and improved patient experiences are clear. Yet, as we navigate this new landscape, it is essential to maintain a focus on ethical considerations and public trust. The future of dentistry, enhanced by robotics and AI, holds the promise of improved oral health for all, provided we proceed with a balanced and responsible approach. The journey forward is one of innovation, responsibility, and a commitment to excellence, ensuring that technological advancements serve to enhance, not replace, the human experience in healthcare.